# Welcome to Your AI-Readiness Playbook!

Adopting AI can seem daunting, especially if you're a small to medium-sized business navigating the fast-moving world of technology.

But the truth is, **AI is more accessible than ever,** and it's not just a tool for large enterprises anymore. In fact, it can be a **game-changer** for SMEs looking to streamline operations, improve customer experiences, and stay ahead of the competition.

**This playbook is designed to help you assess where your business stands when it comes to AI-Readiness.**

We'll walk you through essential areas like technology, data, people, and strategy to help you understand your current position and what steps to take next.

Whether you're just starting to explore AI or you've already begun your journey, **this guide will provide you with practical insights and action steps to make the most of what AI can offer your business.**

By the end of this playbook, you'll not only know how ready your business is for AI but also have a clear path forward—whether it's through self-paced learning with our online courses or by booking a personalised consultation to dive deeper into your specific needs.

Let's get started on your AI journey!

CEO - AI Copilot™, Candice DeVille

# Table Of Contents

# Section 1:

# Technology Readiness

## Building the Foundation for AI Success

### How's your tech holding up?

Think of AI as the engine of a high-performance race car. But for that engine to run smoothly, it needs the right parts—a solid chassis, high-quality fuel, and a reliable pit crew.

In the world of business, your technology infrastructure is the chassis that supports the AI engine.

Without it, the engine (AI) may sputter or fail to deliver the results you're looking for.

In this section, we'll take a look under the hood of your current technology setup and see if it's ready to handle AI.

Don't worry if you're not sure where you stand—by the end of this, you'll have a clear idea of what needs fine-tuning.

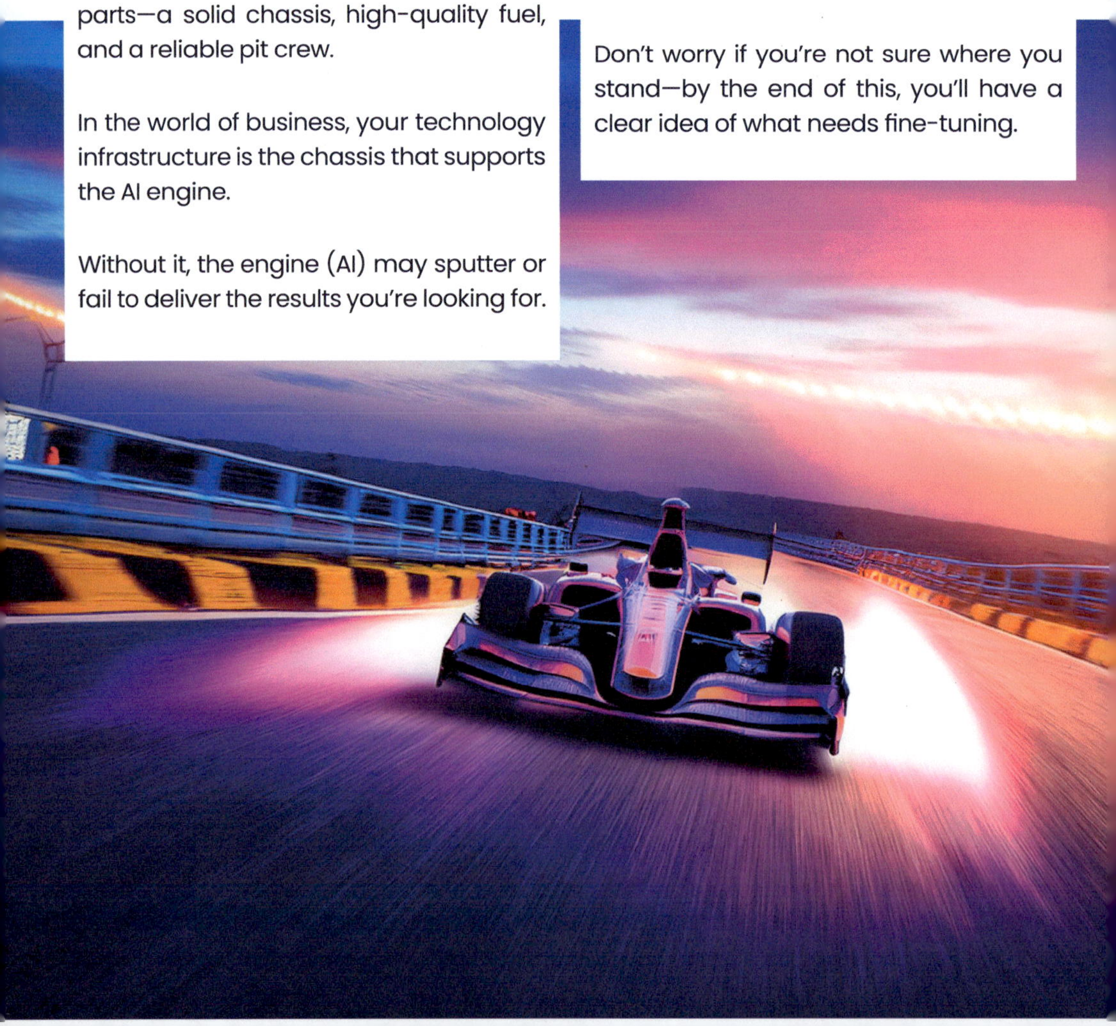

# Why is Technology Readiness important for AI?

Let's start with the basics: for AI to work its magic, it needs a few key ingredients—**data, computing power, and integration.**

If your business is relying on outdated systems, AI may struggle to perform efficiently, like trying to run a modern app on a phone from 2010.

Here's why it matters:

- **Data is King**
  AI thrives on data, and without reliable infrastructure to store and process that data, you won't be able to harness its full potential.

- **Cloud vs. On-Premise**
  Many modern AI solutions work best with cloud-based systems because they allow for scalability and flexibility. If your business is still using old servers, it might be time to consider upgrading.

- **Integration is Key**
  Your systems should be able to "talk" to one another. For example, if your CRM isn't integrated with your sales platform, AI may not have the full picture when analysing customer data.

# Cloud-Based Infrastructure for SMEs: What It Looks Like

When we talk about **cloud-based infrastructure** for small businesses, we aren't just referring to high-end data centres or complex cloud computing systems.

For most SMEs, this means using **cloud platforms** like **Google Workspace** (formerly G Suite), **Microsoft 365**, or other similar tools that allow them to store, access, and manage data online without the need for physical servers.

For many small businesses, being "cloud-based" means using platforms that allow employees to collaborate, store data, and run essential business functions **online.**

Platforms like **Google Workspace** (which includes Gmail, Google Drive, Google Docs, Google Sheets, etc.) or **Microsoft 365** are perfect examples.

# HOW DO CLOUD-BASED PLATFORMS IMPROVE SMALL BUSINESS OPERATIONS?

## 1. Data Storage & Access

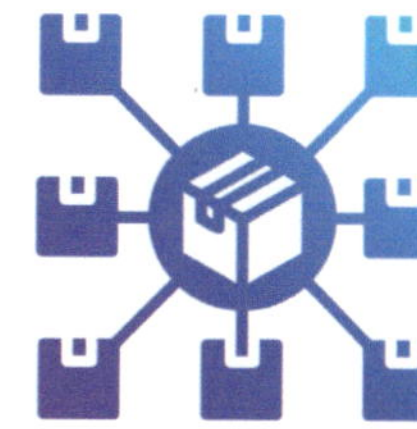

**In Google Drive or OneDrive,** businesses store documents, spreadsheets, presentations, and other critical files. Since everything is stored in the cloud, employees can access this data from anywhere, using any device with an internet connection. **This eliminates the need for physical servers on-site.**

**Why It Matters for AI:** When data is stored in the cloud, AI tools can access and process that information more easily, allowing for faster implementation of machine learning models, automated insights, and other AI-driven processes.

## 2. Collaboration & Communication

**Tools like Google Docs, Google Sheets, Microsoft Teams, and Outlook** make collaboration seamless. This means teams can work on the same document simultaneously, hold virtual meetings, and share data instantly.

**Why It Matters for AI:** AI thrives on real-time data and fast communication. With a cloud-based infrastructure, AI tools can be integrated into your daily workflows—automating tasks like email responses, calendar scheduling, or even helping generate reports in real-time.

## 3. Scalability

**Google Workspace** and similar platforms are highly scalable. Whether you're a team of 5 or 500, you can easily add users, increase storage, or integrate new tools as your business grows—without the need for purchasing new hardware.

**Why It Matters for AI:** AI often requires businesses to scale data processing capabilities. With cloud platforms, you can quickly expand your capacity to handle more data as your AI initiatives grow.

## 4. Affordability

Smaller businesses typically don't have the budget for dedicated servers or complex infrastructure. **Tools like Google Workspace** provide affordable, subscription-based access to cloud storage, communication tools, and collaborative features, which are critical for day-to-day operations.

**Why It Matters for AI:** Affordable cloud solutions mean that even small businesses can implement AI tools without heavy upfront investment in physical infrastructure.

# Google Workspace as Cloud-Based Infrastructure

When we say a small business is ready for AI with **cloud-based infrastructure,** we mean platforms like **Google Workspace** are already in place, offering:

- **Centralised data storage** (Google Drive)
- **Real-time collaboration and communication** (Google Docs, Sheets, Gmail)
- **Scalable and flexible solutions** that allow for growth as AI needs increase
- **Integration-ready platforms** where AI tools (e.g., Google Cloud's AI/ML solutions) can plug in and start processing data with minimal setup.

# Quick Checklist:

## How Ready Is Your Tech?

Here are some questions to help you assess your current technology readiness.

Don't worry if you're not fully there yet—this is about learning where you stand and what steps you can take next.

1. Do you have cloud-based infrastructure or modern servers?

   ☐ Cloud platforms (e.g., AWS, Google Cloud) make it easy to scale AI solutions as your needs grow.

   **Why it matters:** Think of the cloud as a flexible workspace that can expand or contract as your AI needs change. Without it, you might hit a wall when trying to scale your AI initiatives.

2. Is your data structured and easily accessible?

   ☐ Are you storing your data in a way that AI can easily access and analyse it?

   **Why it matters:** Data is the fuel for AI. If your data is disorganised or trapped in silos, your AI won't be able to make accurate predictions or decisions.

3. Are your systems integrated?

   ☐ Do your sales, marketing, customer service, and operational platforms communicate with one another?

   **Why it matters:** Integration is like the wiring in your car—it connects everything and makes sure the right signals get to the right places. Without it, AI will struggle to work across your business effectively.

4. Do you have the bandwidth for AI processing?

   ☐ How much computing power do you have at your disposal? AI can be resource-heavy, so it's important to ensure your systems can handle it.

   **Why it matters:** AI requires computing power to process large datasets, run algorithms, and deliver insights. If your system's too slow, AI could take ages to give you results.

# What If You're Not Ready Yet?

No worries! Many SMEs start out with a few gaps in their tech setup, and that's okay. Here's what you can do to start moving in the right direction:

**UPGRADE IN STEPS**
Consider moving key systems to the cloud over time. This can often be done incrementally without major disruptions.

**FOCUS ON DATA FIRST**
Start by cleaning up your data and setting up a central system to store and organise it.

**EVALUATE YOUR INTEGRATION**
Work with your IT team (or a consultant) to ensure that your systems are talking to each other. A basic integration today can open doors for AI tomorrow.

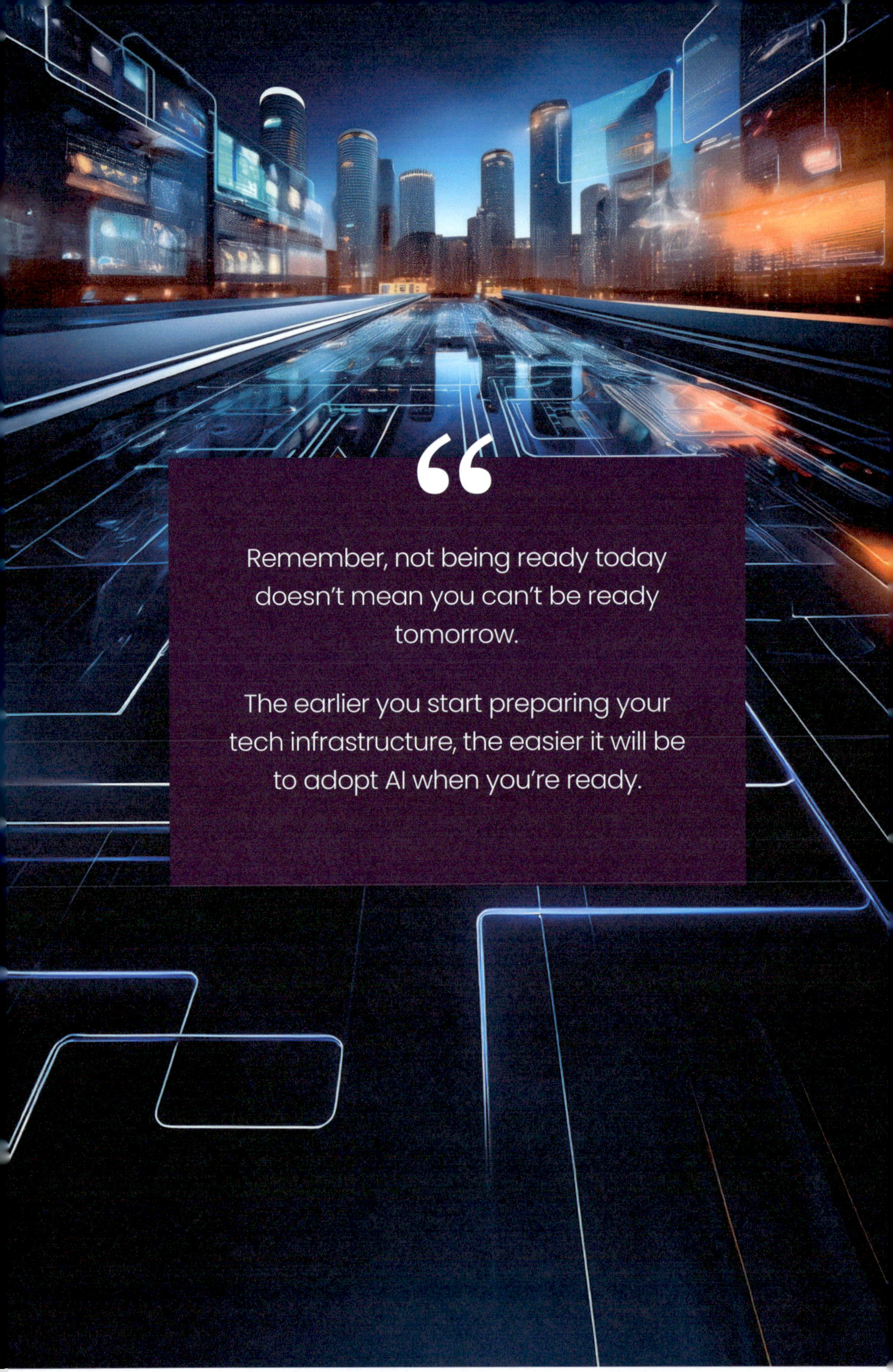

"

Remember, not being ready today doesn't mean you can't be ready tomorrow.

The earlier you start preparing your tech infrastructure, the easier it will be to adopt AI when you're ready.

# Real world client examples:
## **Transforming Cashflow and Efficiency with Artificial Intelligence**

### PROBLEM:

A Managed Service business faced a significant challenge with overdue debtors, straining their cashflow and operational efficiency.

Their CFO was spending an additional full day each week chasing payments, yet overdue debts frequently extended to 120 days or more.

### AI-DRIVEN SOLUTION:

To address this, we collaborated to design and implement an AI-powered system that revolutionised their debt management process.

**The results were transformative: overdue debts were reduced by over 70%, freeing up a full day per week for the CFO.**

This improvement not only optimised cashflow but also enhanced efficiency, strengthened customer relationships, and reduced overhead costs.

This is the power of AI in action—**delivering measurable impact where it matters most.**

# Conclusion:
# Moving Forward

By now, you should have a clearer picture of where your technology stands.

If you've checked off all the boxes, fantastic—you're on the right path. If not, no worries! This is your chance to build a foundation that will support your AI-driven future.

In the next section, we'll dive into your **data readiness**—because without organised and structured data, even the best tech infrastructure won't get you far.

# Section 2:
# Why Data is King
## Unlocking AI's Potential with Proper Data Management

In today's world, **data** isn't just a buzzword—it's the backbone of every business decision, process, and strategy. But when we talk about data, we don't just mean large databases or analytics reports.

We're talking about **everything your business produces and collects**—from meeting notes to customer feedback, internal documents, and operational spreadsheets.

All of this information forms the lifeblood that powers AI and machine learning tools.

Let's break down exactly what we mean by data and how businesses can store and manage it effectively to unlock the full potential of AI.

# What Do We Mean by Data?

At its core, **data** encompasses everything your business uses to function on a day-to-day basis. This includes:

- **Customer Information:**
  Names, contact details, purchasing behaviour, customer preferences, and feedback collected from interactions or surveys.

- **Operational Data:**
  Sales reports, supply chain information, financial statements, and inventory levels that keep the business running smoothly.

- **Internal Documents:**
  Meeting notes, project plans, strategy documents, and HR files. These include everything from employee contracts to performance reviews.

- **Emails and Communications:**
  Internal communications between teams, customer support emails, and even conversations on platforms like Slack or Microsoft Teams.

- **Website Data:**
  Information collected from website traffic, user behaviour, and conversion metrics.

- **Marketing Data:**
  Campaign performance, ad spend, click-through rates, and lead generation stats.

"

In essence, every single interaction your business has—internally and externally—generates data.

Storing and managing this data effectively is crucial to allow AI systems to sift through it and deliver insights.

# How Businesses Store Data: The Fundamentals

## 1. Cloud Storage

**What it is:** Cloud-based storage solutions (e.g., Google Drive, AWS, or Microsoft Azure) allow businesses to store data remotely on servers that can scale based on demand.

**Why it's useful:** It offers flexibility and scalability, meaning as your business grows, your storage can grow with you. Cloud storage also makes data more accessible, which is key for AI tools that need to retrieve large amounts of data from different parts of the business quickly.

## 2. On-Premise Servers

**What it is:** Some businesses still use physical servers located in-house to store data.

**Why it's useful:** This provides direct control over security and access but can be limiting in terms of scalability. It also requires ongoing maintenance and physical space, which isn't always feasible for SMEs.

## 3. Data Lakes and Warehouses

**What it is:** These are large-scale storage repositories that can hold structured and unstructured data. A **data lake** can store vast amounts of raw data in its native format, while a **data warehouse** stores structured and organised data for easier analysis.

**Why it's useful:** Data lakes are ideal for AI because they allow businesses to store everything in one place, without the need to organise it right away. Once ready, data can be processed and analysed in a more structured way using a warehouse or other tools.

# Why Data is King: Unlocking AI's Potential with Everyday Business Data

For AI to be effective in your business, it needs access to all the information you generate on a daily basis—yes, all of it!

This means not only the obvious stuff like customer records and sales reports, but also the meeting notes, documents, and even internal communications that help run your business.

Most SMEs today operate within ecosystems like **Microsoft 365 or Google Workspace** to manage their day-to-day tasks, and these platforms generate and store massive amounts of valuable data.

Let's look at exactly how your business might be collecting and storing data daily—and how it can all be used by AI to transform your operations.

# How Businesses Collect and Store Data Day-to-Day

## 1. Emails and Communications

### (e.g., Microsoft Outlook, Gmail)

Every day, businesses send and receive emails containing critical information: project updates, customer inquiries, internal team discussions, and more.

This communication data is often stored automatically in platforms like Outlook or Gmail.

**Example:** Every time you receive an email from a customer with a support question or feedback, this is valuable data that can be analysed by AI to improve customer service or identify trends in customer needs.

**AI Application:** AI tools can analyse email patterns and flag common issues that could improve customer satisfaction, or even automate routine responses.

## 2. Internal Documents and Files

### (e.g., OneDrive, Google Drive)

Businesses store everything from contracts and strategy documents to meeting minutes and marketing plans on cloud platforms like **OneDrive (Microsoft) or Google Drive (Google Workspace).**

These documents are critical for day-to-day decision-making and operational efficiency.

**Example:** Every time your team creates a meeting agenda or a report, it's uploaded to the cloud, making it easily accessible for future reference.

AI can scan these documents for patterns, such as recurring action items or common project delays.

**AI Application:** AI can automatically sort and tag documents for easier access or even summarise the content of large documents for quick reference, helping your team be more efficient.

## 3. Collaboration Tools

### (e.g., Microsoft Teams, Google Meet)

Teams are increasingly relying on collaboration platforms like **Microsoft Teams or Google Meet** to communicate and manage projects.

Every conversation, shared document, and video meeting contributes to the business's data pool.

**Example:** Think about a weekly team meeting where project milestones are discussed.

The notes taken during the meeting are valuable data that can be analysed by AI to highlight critical decisions or track progress over time.

**AI Application:** AI can process meeting notes and conversations, automatically summarising key points, assigning tasks, or even suggesting follow-up actions based on past meetings.

## 4. Customer Interactions

### (e.g., CRM integrations with Microsoft Dynamics or Google Sheets)

Customer relationship management (CRM) tools integrated with platforms like **Microsoft Dynamics or Google Sheets** allow businesses to track sales, customer interactions, and service requests.

This data is crucial for understanding customer behaviour and preferences.

**Example:** Each time a sales rep logs a customer call or updates a lead's status in your CRM, it's data that AI can analyse to predict which customers are likely to convert, or even recommend which actions to take next to close a sale.

**AI Application:** AI can analyse customer interactions to identify patterns, like which touchpoints lead to conversions or where customer service bottlenecks occur.

## 5. Project Management Tools

### (e.g., Microsoft Planner, Google Keep)

Many businesses manage their projects using tools like **Microsoft Planner or Google Keep** to track tasks, deadlines, and team responsibilities.

Every task assigned or note added contributes to your overall data landscape.

**Example:** If your team uses Microsoft Planner to track the progress of a marketing campaign, every task completion and status update is data that AI can use to analyse project efficiency and make suggestions for improving workflows.

**AI Application:** AI can predict project delays based on task completion patterns and suggest adjustments to keep everything on track.

## 6. Spreadsheets & Financial Data

## (e.g., Microsoft Excel, Google Sheets)

Most businesses use spreadsheets to track finances, budgets, inventory, and more.

Platforms like **Excel and Google Sheets** are often the go-to tools for organising this information.

**Example:** Weekly financial reports or inventory updates entered into a shared Google Sheet become a treasure trove of data.

AI can analyse this data to optimise stock levels, detect fraud, or forecast future sales.

**AI Application:** AI tools can take your financial data and make forecasts, detect patterns in cash flow, or even automate repetitive tasks like generating monthly reports.

## 7. Customer Feedback & Surveys

## (e.g., Microsoft Forms, Google Forms)

Businesses often collect customer feedback through forms or surveys created in **Microsoft Forms or Google Forms.**

This feedback, whether it's product reviews or customer satisfaction scores, is a valuable source of data for improving business operations.

**Example:** After launching a new product, you send out a customer satisfaction survey.

The responses, which include both quantitative ratings and qualitative comments, provide data for AI to analyse customer sentiment.

**AI Application:** AI can scan customer feedback to identify recurring issues or trends and automatically generate insights about customer sentiment that could guide future product improvements.

# Making Your Data Work for You

So, what does all this data mean for AI? Simply put, the better your data is stored and managed, the more valuable insights AI can deliver.

AI can help you spot trends, automate processes, and make smarter decisions—but only if your data is accessible, structured, and secure.

### Centralization

Start by centralising your data storage.

Whether you're using Google Drive or Microsoft OneDrive, aim to keep everything in one place to make it easier for AI tools to access and analyse.

### Consistency

Make sure data entry is consistent across your platforms.

For instance, ensure that customer names, dates, and figures follow the same format in every document and system.

### Security

Prioritise data security and privacy.

Cloud platforms like Microsoft 365 and Google Workspace have built-in security features, but you'll also want to implement access controls and regular audits to ensure your data stays safe.

# Why Data Matters for AI?

AI systems thrive on **data** because that's where they learn. The more data an AI tool has access to, the more accurately it can perform tasks like making predictions, automating processes, and delivering insights.

Here's why the way you store your data is crucial for AI:

1. **AI Needs Large, Quality Data Sets**

2. **Structured vs. Unstructured Data**

3. **Data Accessibility**

4. **Security and Privacy**

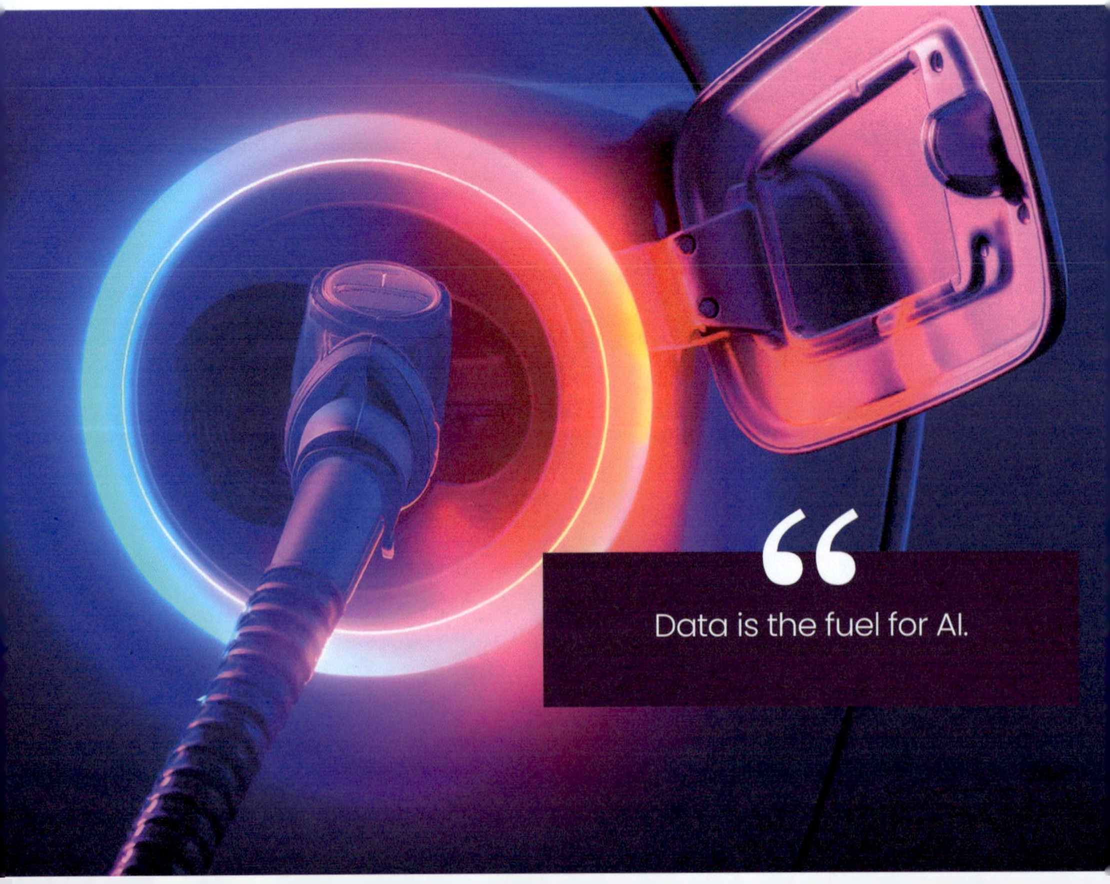

## 1. AI Needs Large, Quality Data Sets

For AI to deliver useful insights, it needs access to large volumes of data. If your data is scattered across multiple platforms or stored in formats that AI can't access, its performance will be severely limited.

Imagine trying to assemble a puzzle with only half the pieces—it's the same for AI with missing or incomplete data.

## 2. Structured vs. Unstructured Data

**Structured data (like numbers in a spreadsheet)** is easy for AI to process, while **unstructured data (like emails, PDFs, or meeting notes)** requires more advanced AI algorithms to analyse. Having both types organised and accessible helps AI deliver more accurate results.

For example, AI can analyse customer emails alongside sales reports to detect patterns in customer feedback and behaviour.

## 3. Data Accessibility

AI needs to be able to retrieve and process data in real-time. If your data is locked away in silos—separate systems that don't communicate with each other—AI won't be able to deliver the insights you're hoping for.

Cloud storage and system integration are crucial to ensure all the necessary data is accessible.

## 4. Security and Privacy

Handling data responsibly is vital, especially with personal or sensitive information.

AI can only be as good as the data it's trained on, so businesses need to ensure that they're storing data securely, adhering to privacy laws like GDPR, and being mindful of how they collect and manage customer information.

# Action Steps:

## How to improve Data Storage and Readiness for AI

### Centralise Your Data:

Move from scattered spreadsheets and disconnected tools to a centralised data storage solution, like a **data lake or cloud platform (eg: Google Workspace, AWS).** This will make it easier for AI to access and analyse everything in one place.

### Ensure Data Consistency:

Standardise how you store different types of data (e.g., using consistent formats for customer names, order numbers, or dates). This makes it easier for AI to process and use the data effectively.

### Back-Up and Secure:

Regularly back up your data to avoid losing valuable information. Make sure you're also protecting sensitive data with encryption and access controls to maintain compliance with privacy laws.

### Integrate Your Systems:

Ensure your CRM, marketing platforms, sales tools, and financial systems are all connected. This integration allows AI to pull data from all parts of your business for a more comprehensive analysis.

# Real world client examples:

## Revolutionising Customer Support with Artificial Intelligence

### PROBLEM:

A large national infrastructure company was overwhelmed by the growing volume of inquiries and support tickets.

They needed a scalable, reliable, and 24/7 solution that could deliver accurate, multi-lingual responses without adding costly infrastructure or overhead.

### AI-DRIVEN SOLUTION:

We partnered with the company to develop a custom AI Agent chat system tailored to their needs.

**The AI seamlessly recognised the user's input language and responded in kind, ensuring a smooth and personalised experience for every customer.**

This innovative solution not only eliminated **the need for additional resources but also significantly enhanced the customer experience**.

Moreover, the AI Agent's built-in analytics provided valuable insights into customer sentiments, frequently asked questions, and emerging trends, empowering the business to make data-driven decisions and guide future developments.

With this AI-driven transformation, **the company now delivers exceptional support while staying agile and efficient in a competitive landscape.**

# Conclusion:
# Data as the Fuel for AI

Every email you send, spreadsheet you update, and meeting note you take is a small piece of a much larger puzzle.

AI can help put that puzzle together, but only if the pieces are well-organised and accessible.

By centralising and managing your day-to-day data more effectively, you'll be setting your business up to take full advantage of what AI can offer.

In the next section, we'll look at **People and Skills Readiness**—because technology alone won't drive AI success. You need the right team behind it too!

# Section 3:
# People & Skills Readiness
## Building the Right Team for AI Success

When it comes to adopting AI, the people behind the technology are just as important as the technology itself.

AI can automate tasks, analyse data, and provide insights, but it's your team that will decide **how** to implement those insights, adjust business strategies, and drive innovation.

In this section, we'll explore why human skills are critical for AI success and how to ensure your team is ready to make the most of AI tools.

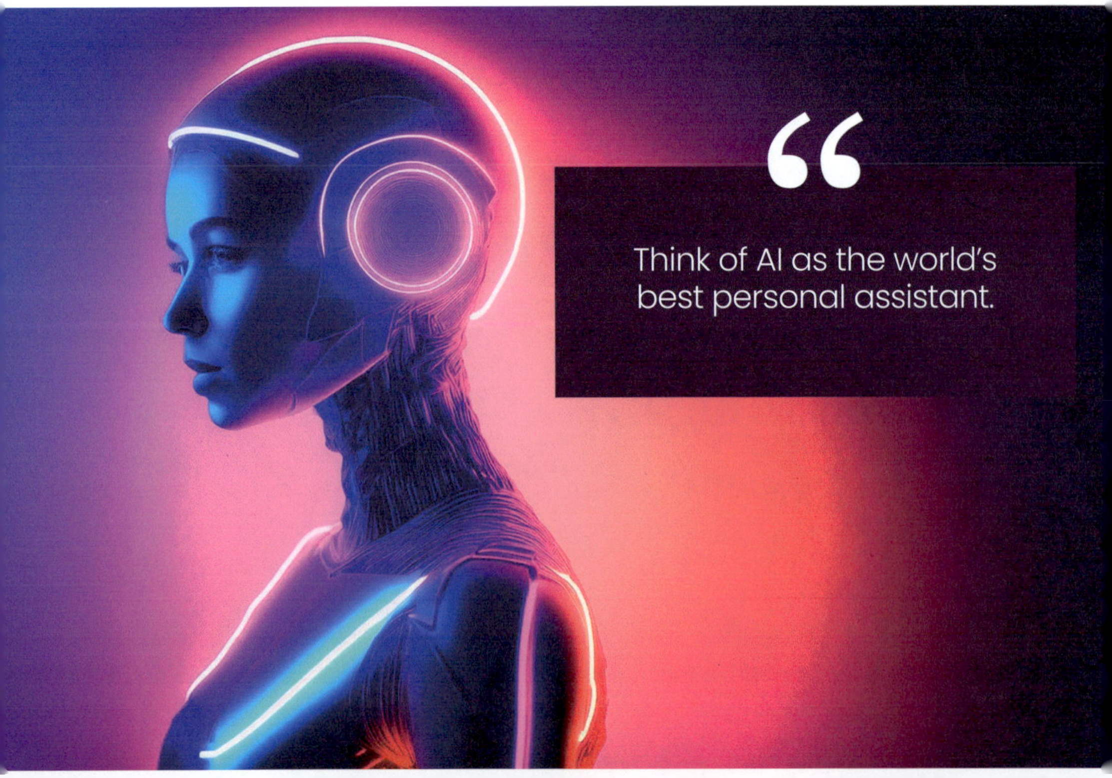

## Why Human Skills Matter for AI Success?

AI is powerful, but it's not a one-size-fits-all solution. To truly succeed with AI, your business needs people who:

1. Understand how to work alongside AI tools.

2. Interpret AI-driven insights and apply them to business decisions.

3. Ensure AI is implemented ethically and responsibly, avoiding biases or errors in data processing.

AI can handle repetitive tasks, analyse huge volumes of information, and provide recommendations, but it still relies on **you and your team** to steer the ship.

# Key Skills Needed for AI Adoption

Let's look at the specific skills your team needs to be AI-ready.

These skills can be divided into two main areas: **technical skills and soft skills.**

**TECHNICAL SKILLS:**

Data Literacy
Basic AI Literacy
Data Analysis
IT Skills for System Integration

**SOFT SKILLS:**

Adaptability
Problem Solving
Communication

# TECHNICAL SKILLS

Even if your team doesn't need to build AI tools from scratch, having a solid understanding of how AI works—and how to manage and interpret data—will be critical.

## Data Literacy:

Your team needs to understand how to work with data, even at a basic level. This means knowing how data is collected, stored, and used, and how to interpret insights generated by AI systems.

**Example:** Let's say your AI tool provides a customer behaviour analysis based on sales data. Someone on your team should be able to assess whether the data is reliable, spot trends, and make recommendations based on the insights provided.

## Basic AI Literacy:

It's not necessary for your team to become AI experts, but they should understand **what AI can and can't do.** Knowing AI's capabilities helps set realistic expectations and ensures the technology is used appropriately.

**Example:** Your marketing team should know that while AI can help predict customer trends, it might not replace the need for human creativity in campaign development.

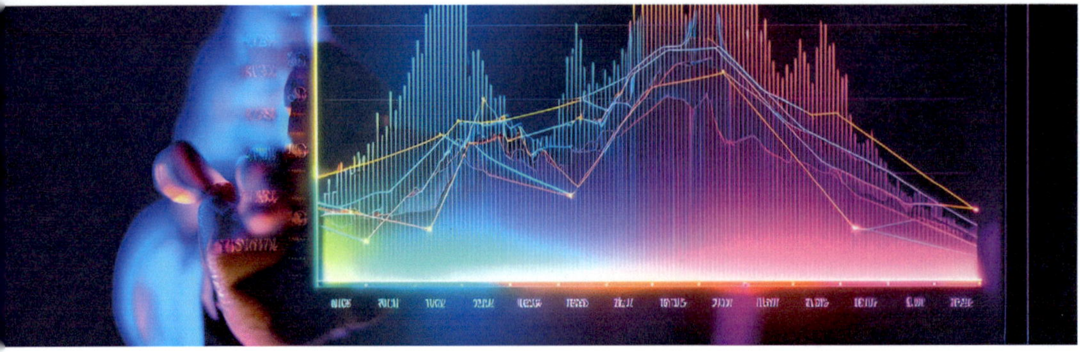

## Data Analysis:

Having team members who can dive deeper into the data, identify patterns, and use AI-generated insights for better decision-making will be a game-changer. Skills like working with tools like Excel, Google Sheets, or more advanced BI platforms **(like Power BI or Google Data Studio)** are important.

**Example:** After running a promotional campaign, an employee with data analysis skills can use AI tools to evaluate which customer segments were most engaged and adjust future campaigns accordingly.

## IT Skills for System Integration:

Your IT team will need to ensure that AI systems are integrated smoothly with your existing platforms (CRM, ERP, etc.). This requires an understanding of both your current infrastructure and how AI tools fit in.

**Example:** If you're using Microsoft 365, your IT team should ensure that AI systems like Microsoft Power Automate or Azure AI are properly integrated and pulling data from the right sources.

# SOFT SKILLS

Along with technical abilities, certain soft skills are just as crucial for a successful AI transformation.

## Adaptability:

AI adoption brings change, and not everyone is comfortable with change. Teams that are adaptable and open to learning new tools and processes will thrive.

**Example:** If AI is implemented to streamline customer service tasks, the customer service team should be flexible enough to adjust to new processes and focus on higher-level inquiries while AI handles routine requests.

## Problem-Solving:

AI tools can provide insights, but your team needs the problem-solving skills to **interpret those insights** and apply them to real-world challenges.

**Example:** AI might suggest that your inventory management system is inefficient, but it's your team's problem-solving abilities that will identify which specific actions to take to address the issue.

## Communication:

As AI systems generate new insights and recommendations, your team will need to effectively communicate those findings across the business. Being able to explain AI-driven insights to non-technical team members will ensure everyone is on the same page.

**Example:** Your marketing manager may need to explain to your sales team how AI insights from customer data can help refine sales pitches or improve customer engagement strategies.

# Action Steps:
## Preparing Your Team for AI

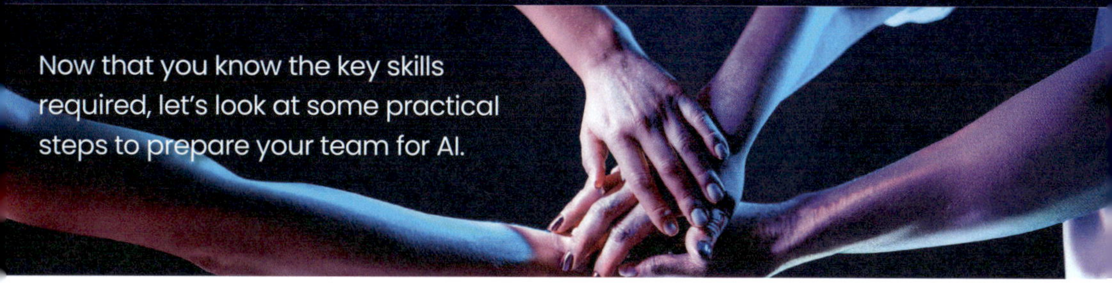

Now that you know the key skills required, let's look at some practical steps to prepare your team for AI.

### Upskill and Reskill:
Encourage team members to develop **basic data and AI literacy** through online courses or in-house training. **Platforms like Coursera, LinkedIn Learning, or Udemy** offer courses on data analysis and here at AI Copilot, we provide courses on AI fundamentals that are accessible to all skill levels.

### Hire AI Expertise Where Needed:
If your team doesn't have the necessary expertise in-house, consider bringing in external consultants or hiring individuals with AI-specific skills to lead the charge

### Encourage Cross-Department Collaboration:
AI touches many areas of your business, from marketing to operations. Ensure that your team members from different departments are communicating effectively and working together to implement AI tools and interpret the results.

### Start Small:
Rather than overwhelming your team with complex AI projects, start with simple tools **(like automated chatbots or data visualisation tools)** that are easier to manage and understand.

# Identifying and Building Internal AI Champions

When introducing AI into your business, you don't just need tech-savvy employees—you need **AI champions.**

These are individuals who will advocate for AI, help integrate it into workflows, and guide others through the transition.

What's more, **AI champions aren't limited to the IT department.** Employees with strong communication, problem-solving, and leadership abilities, regardless of their department, can become powerful AI advocates.

**Here's how businesses can identify and nurture AI champions, and the soft skills to look for.**

**Step 1: Identifying Potential AI Champions**

**Step 2: What Soft Skills to Look For in AI Champions**

**Step 3: Building and Nurturing AI Champions**

> “While technical skills are important, soft skills play a key role in building these AI champions.

# Step 1:
## Identifying Potential AI Champions

AI champions can come from unexpected places within your business. Don't assume that only your IT or data team can take on this role.

The best AI champions often combine enthusiasm for new technologies with strong leadership and communication skills. Here's how to spot them:

### 1. Look Beyond The Usual Suspects

**Not just IT:** AI champions don't have to be experts in AI itself. While technical know-how is useful, the real magic comes from employees who can bridge the gap between AI's capabilities and practical business applications.

**Example: A customer service manager with a passion for improving customer experiences might be a perfect AI champion to implement a chatbot or automate customer queries.**

**Their deep understanding of customer needs can guide AI implementations in meaningful ways.**

## 2. Gauge Interest and Enthusiasm

Employees who are naturally **curious** about new technologies, constantly asking questions about how things can be done more efficiently, or suggesting improvements in processes are great candidates.

Look for those who've shown interest in learning about AI, data, or automation—even if they're not yet experts.

**Example: A marketing team member who's enthusiastic about using AI for personalised campaigns or analytics might be well-suited to spearhead AI adoption within their department.**

**Their deep understanding of customer needs can guide AI implementations in meaningful ways.**

## 3. Observe Cross-Functional Collaboration

Potential AI champions are often **natural collaborators**—those who work well with other departments and understand how AI might solve broader business challenges, not just those within their niche.

**Example: An operations manager who frequently collaborates with the sales team might see opportunities for AI to streamline both processes, making them a valuable advocate for adopting AI across multiple functions.**

# Step 2:
## What Soft Skills to Look For in AI Champions

AI champions need to have a combination of **leadership, communication, and problem-solving skills** that will allow them to inspire others and drive AI adoption effectively.

Here are te top soft skills to focus on when selecting AI champions:

### 1. Amazing Communication Skills

One of the most critical skills for an AI champion is the ability to **translate complex AI concepts into everyday language.** They need to communicate the benefits and practical applications of AI to colleagues who may not be familiar with the technology.

**Example: A marketing coordinator who can explain how AI-driven customer segmentation works without diving into data science jargon could help their team embrace AI's benefits more easily.**

## 2. Problem-Solving Mindset

AI champions should be natural **problem solvers** who look at business challenges and see AI as a tool to solve them. They should be able to connect the dots between business problems and AI solutions.

**Example: A sales manager who notices bottlenecks in lead management might recognize how an AI-powered CRM can help automate follow-ups and prioritise leads, improving conversion rates.**

## 3. Empathy and Team Orientation

Introducing AI can be disruptive, and not everyone will be on board right away. AI champions need to have **empathy and the ability to support colleagues** through change. This includes addressing fears around job displacement and showing how AI can enhance, rather than replace, human roles.

**Example: An HR professional who can empathise with employees' concerns about AI but also highlight how AI can reduce repetitive tasks and free up time for more meaningful work could make an ideal AI champion.**

## 4. Adaptability and Openness to Learning

AI is constantly evolving, and your champions need to be **open to learning** and adapting as new tools and technologies emerge. They should be comfortable experimenting with AI, even if they don't have all the answers right away.

**Example: A product manager who is always on the lookout for the latest trends and is open to experimenting with AI for product development or customer feedback analysis will be able to guide AI projects with agility.**

## 5. Leadership by Influence, Not Authority

AI champions might not always hold formal leadership roles, but they should have the ability to **influence their peers** and inspire others to adopt AI. They lead by example, showing how AI can solve problems and improve workflows.

**Example: A mid-level analyst who constantly finds better ways to use data and gets their colleagues excited about using AI tools can become an influential AI champion within the organisation.**

# Step 3:

## Building and Nurturing AI Champions

Once you've identified potential AI champions, the next step is to **support and nurture** their growth. Here are some strategies for helping them thrive:

### 1. Provide Learning Opportunities

Encourage your AI champions to **take the lead on learning** by offering them access to AI courses, certifications, or workshops. At AI Copilot, we offer practical AI courses specifically tailored for business applications.

**Action: Set up internal workshops or sponsor external training to deepen their knowledge of AI tools relevant to your business.**

### 2. Encourage Cross-Department Collaboration

Have your AI champions lead **cross-functional teams** that include IT, marketing, HR, and operations. This helps ensure that AI is understood and applied across the entire business, not just one department.

**Action: Give AI champions opportunities to present their insights at company-wide meetings or cross-team workshops, showcasing practical applications of AI.**

## 3. Foster a Culture of Experimentation

Allow your AI champions to **experiment** with small AI projects before implementing larger-scale solutions. Giving them the freedom to try new tools and processes in a low-risk environment builds confidence and momentum.

**Action: Start with pilot projects that give AI champions room to test solutions (e.g., an AI chatbot for customer service or AI-based marketing automation).**

## 4. Recognize and Reward

AI Acknowledge your AI champions' contributions, both formally and informally. Recognition motivates them to continue leading AI efforts and encourages others to get involved.

**Action: Highlight their successes in company newsletters, team meetings, or through internal awards programs focused on innovation.**

# Real world client examples:
## Accelerating Sales with AI-Driven Recommendations and Quoting

### PROBLEM:

A large product wholesaler sought a solution to streamline their sales process by recommending and quoting products, alleviating the burden on their sales team.

They needed an efficient, scalable system capable of handling complex and specific customer requirements.

### AI-DRIVEN SOLUTION:

We developed a cutting-edge AI agent that intelligently recommended and quoted products based on precise criteria.

**The results were remarkable: the sales team's workload was reduced by over 60%, while the lead-to-close time dropped by an impressive 90%.**

Additionally, the solution boosted the company's sales conversion rate by more than 30%.

This AI-powered transformation not only **empowered the sales team to focus on high-value opportunities but also enhanced overall efficiency and accelerated business growth.**

# Conclusion:
# Building an AI-Ready Team

While technology and data are critical components of AI, it's your **people** who will drive success.

By equipping your team with the right skills—both technical and soft—you'll create a workforce that's not only capable of using AI but one that's ready to leverage it to grow your business.

In the next section, we'll look at **Business Readiness** and how to align AI with your overall business goals and strategy.

# Section 4:
# Business Readiness
## Aligning AI with Your Business Goals

Adopting AI isn't just about having the right technology or data; it's about **aligning AI with your broader business goals.**

AI should serve as a tool to enhance and grow your business, helping you solve real-world challenges, improve efficiency, and scale your operations.

In this section, we'll explore how businesses can assess their overall readiness to implement AI by ensuring that the technology supports their **strategic objectives** and fits into their long-term plans.

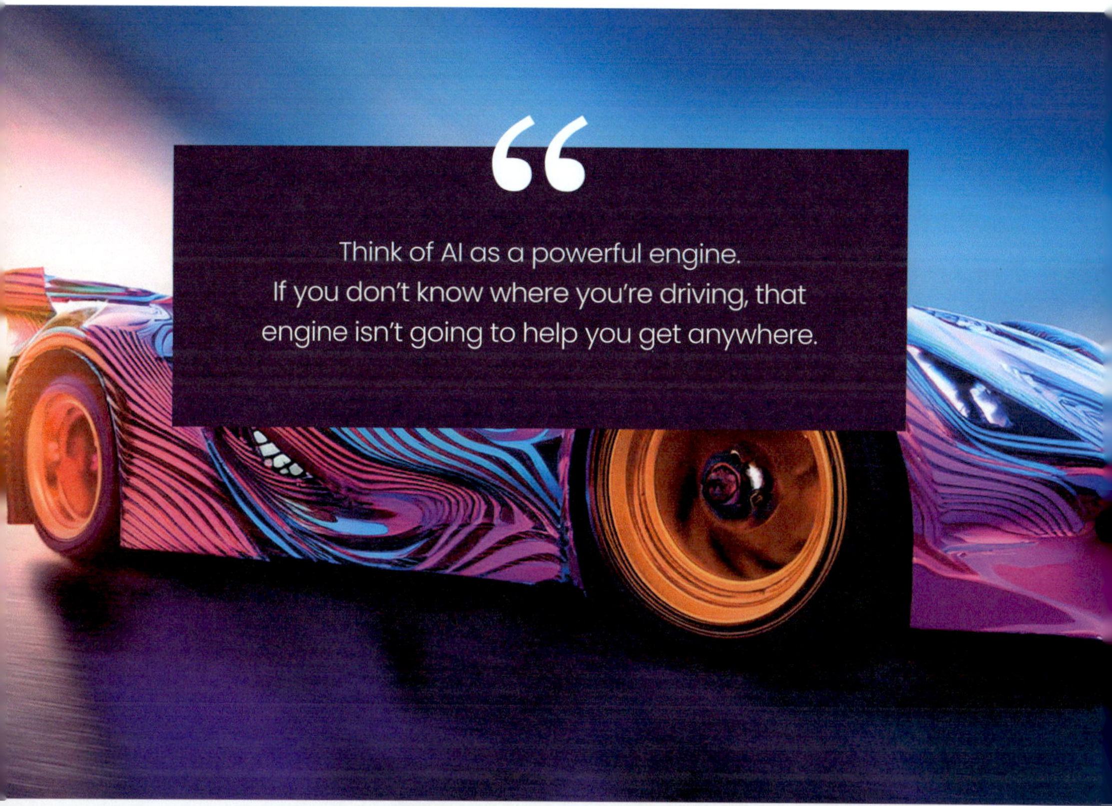

## Why Business Readiness Is Crucial for AI Adoption

Implementing AI successfully requires a clear **business case.** Without aligning AI efforts with the specific goals of your business, you run the risk of investing in technology that looks impressive but doesn't deliver real results.

In this section, we'll help you figure out where you're headed and how AI can help you get there.

## Key Areas to Assess for Business Readiness

When determining whether your business is ready for AI, it's important to look at several critical areas:

1. **Strategic Alignment**
2. **Understanding Your Pain Points**
3. **Willingness to Invest in AI**
4. **Flexibility and Change Management**

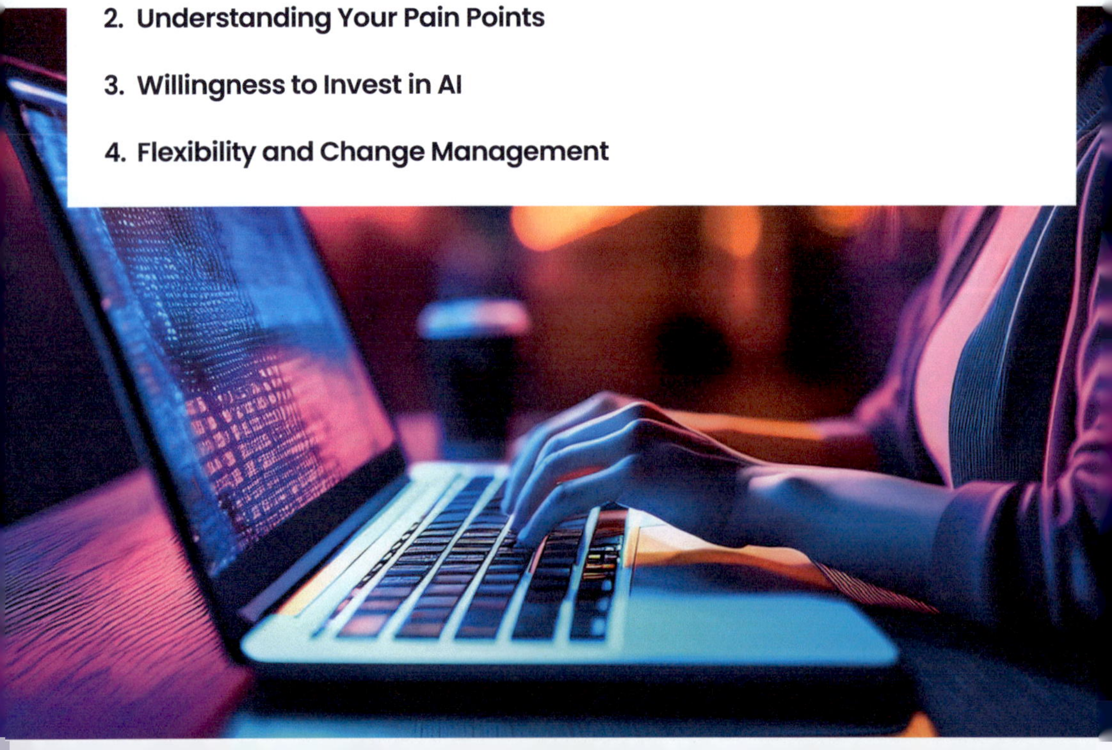

# 1. Strategic Alignment

AI works best when it is used to directly support your **strategic business objectives.** Ask yourself:

### Do you have clear goals that AI can help achieve?

- AI should address specific challenges or opportunities within your business.

  If your goal is to improve customer retention, AI might help by analysing customer behaviour and providing personalised experiences. If your goal is to optimise your supply chain, AI can predict demand and manage inventory.

- **Example: A retail business might use AI to analyse sales trends and recommend which products to stock up on during certain seasons, helping them meet sales goals and reduce overstock.**

### Are you focusing on areas where AI can make a measurable impact?

- Identify high-impact areas where AI can provide tangible results, such as cost savings, increased revenue, or time efficiency.

- **Example: If your sales team spends hours manually updating customer records, automating that process through AI can save time and allow your team to focus on closing more deals.**

# 2. Understanding Your Pain Points

Businesses that successfully implement AI have a clear understanding of the **challenges they are facing.** It's important to identify these pain points and determine if AI can provide solutions.

### Where are you experiencing bottlenecks or inefficiencies?

- If your team is dealing with repetitive tasks, AI can step in to automate them, freeing up valuable time for more strategic activities.

- **Example: An HR team might be bogged down with screening resumes, a process that AI can automate to ensure quicker hiring decisions.**

  **(Note: This type of project requires specific ethical considerations)**

### Do you have specific business processes that could benefit from AI?

- AI thrives in environments where routine processes can be enhanced or automated.

  Look for areas like customer service, marketing automation, or operational logistics where AI can reduce human error and increase productivity.

- **Example: A company struggling to respond quickly to customer inquiries could implement an AI-powered chatbot to handle basic queries 24/7, reducing response times and improving customer satisfaction.**

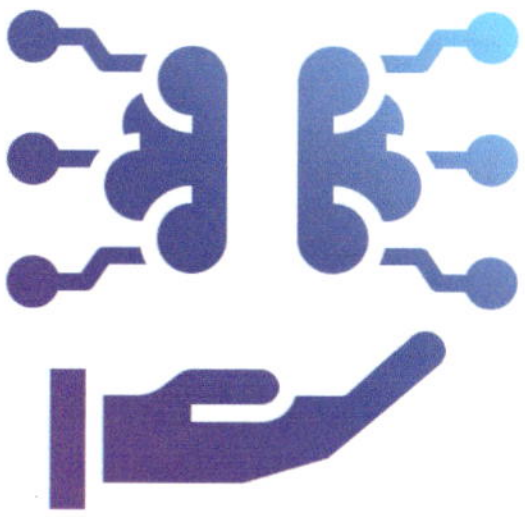

# 3. Willingness to Invest in AI

AI is a long-term investment, and businesses must be ready to commit **time and resources** to see the best results.

### Are you prepared to allocate budget for AI development?

- AI adoption often requires upfront investment, whether it's in hiring external consultants, training employees, or upgrading your technology infrastructure.

  Businesses need to be ready to invest not only in AI tools but also in building the expertise and infrastructure needed to support AI.

- **Example: An SME might start with a small AI-powered marketing automation tool but plan for additional investments as the business scales and sees the benefits.**

### Do you have realistic expectations about ROI?

- AI can deliver powerful results, but businesses must understand that it's not always a quick fix.

  Depending on the complexity of the AI solution and your industry, the return on investment (ROI) might take time to materialise.

- **Example: A manufacturing company investing in predictive maintenance through AI might not see immediate savings but could prevent costly breakdowns in the future, leading to long-term ROI.**

# 4. Flexibility and Change Management

AI adoption often means **changing how your business operates,** and this requires flexibility from both leadership and employees.

### Is your business culture open to adopting new technologies?

- Implementing AI may require shifts in mindset and process.

  It's important to ensure that leadership is on board and that employees are willing to embrace the changes AI brings.

- **Example: A company that promotes innovation and continuous improvement will likely see smoother AI adoption than one where change is met with resistance.**

### Do you have a change management plan?

- Change management is critical when introducing AI.

  Having a clear plan for how to roll out AI tools, train staff, and address any resistance will help smooth the transition.

- **Example: You might roll out AI solutions in stages, starting with smaller, less disruptive areas like marketing automation before expanding to more mission-critical functions.**

# Action Steps:
## Building Business Readiness for AI

Once you've assessed your readiness, it's time to take actionable steps to ensure AI aligns with your business goals:

### Start Small and Scale:

Begin with a pilot AI project in a single department or process where AI can make an immediate impact.

This allows you to test and learn from the process before scaling AI solutions acrss the business.

**Example: A retail business might start with AI-driven personalised marketing campaigns before moving into a larger project like AI for inventory management**

### Develop a Clear AI Strategy:

Your AI efforts should be part of a **larger business strategy.**

Define clear goals for what you want AI to achieve and map out how AI fits into your long-term business plan.

**Example: If your goal is to improve customer retention, your AI strategy could include personalised email campaigns, predictive customer behaviour analysis, and automating customer service responses.**

### Involve Key Stakeholders:

Ensure that leaders from various departments—marketing, sales, operations, and HR—are involved in the AI strategy and implementation process.

Cross-department collaboration ensures AI solutions are integrated effectively across the business.

**Example: Get input from the sales and marketing teams to ensure that any AI-powered CRM tool aligns with their specific needs and workflows.**

> Artificial intelligence is not just about machines; it's about amplifying human intelligence.
>
> — Satya Nadella

# Conclusion:
# Aligning AI with Your Business Goals

AI should be a tool that **enhances your business objectives,** not a solution in search of a problem.

By ensuring that your business is strategically aligned and prepared to invest in AI, you can unlock its full potential and set your company on a path to greater efficiency and growth.

In the next section, we'll cover the **decision matrix,** where you can assess your overall AI-Readiness based on your answers from this and previous sections.

# Introduction to the AI-Readiness Decision Matrix

Before diving into AI, it's important to know just how prepared your business is to implement this powerful technology.

**Our AI-Readiness Decision Matrix** is designed to help you evaluate key areas of your business, including your **technology infrastructure, data management, team skills, and strategic alignment with AI.**

This matrix will guide you through a series of questions that assess where your business currently stands and help you understand areas that might need improvement.

You'll rate your readiness in categories like **technology, risk management, security, and business impact, giving you a clear, actionable score by the end.**

Whether you're on the verge of AI adoption or just getting started, this tool will help you map out your next steps—whether that's diving into training, upgrading systems, or refining your business strategy.

**To check your readiness for AI, use our 3-minute online assessment through QR code below.**

# Congratulations!

## on completing the AI-Readiness Decision Matrix

You've taken the first big step towards AI adoption, and now you have a clearer understanding of where your business stands.

Whether your score shows that you're fully ready or that there's still some work to do, don't worry—there are **actionable next steps** you can take no matter where you are on your AI journey.

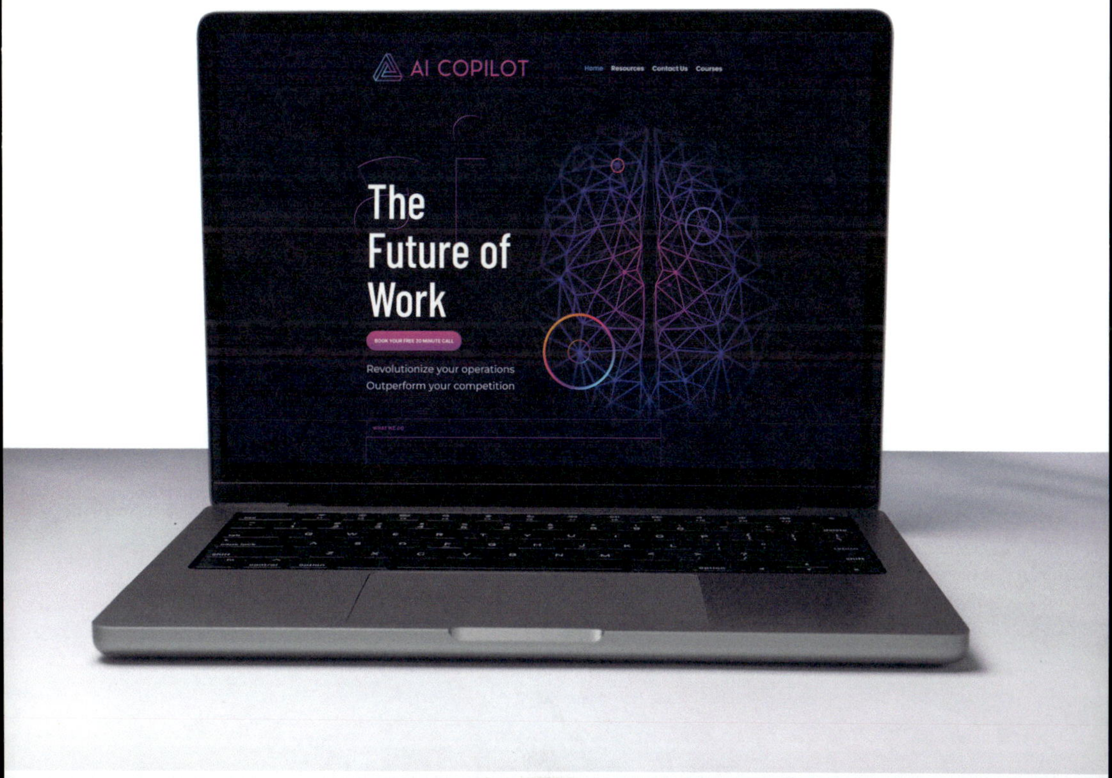

# Next Steps Based on Your Score

## SCORE: YOU'RE READY FOR AI

Fantastic! Now's the perfect time to start exploring the AI tools that align with your business goals. You can dive into areas like automation, data analysis, or even AI-driven customer service.

**What's next?** Book a free consultation to discuss how we can help you integrate AI solutions tailored to your business.

## SCORE: YOU'RE ALMOST READY

You're on the right path but might need a few tweaks to your technology, data management, or team skills. A small investment in these areas can make a big difference.

**What's next?** Check out our upskilling courses that will help your team build the expertise needed to handle AI or schedule a consultation to get personalised advice on fine-tuning your readiness.

## SCORE: YOU'RE NOT QUITE READY

Don't worry! Many businesses start from this point. Focus on improving your infrastructure, organising your data, and developing internal AI champions to lead the way.

**What's next?** Begin with our AI-Readiness Fundamentals course, designed specifically for businesses building their foundation for AI. Or, schedule a free consultation to discuss how to move forward and prioritise your efforts.

“

Even the greatest was once a beginner.
Don’t be afraid to take that first step.

— Muhammad Ali

## START LEARNING:

Dive into our self-paced **online courses** to build the skills and knowledge you need for successful AI adoption.

https://aicopilot.com/learning/

## GET PERSONALISED HELP:

Not sure where to start? Book a **free 30-minute consultation** with one of our AI experts, and we'll help you create a tailored action plan.

https://aicopilot.com

**You're well on your way to harnessing the power of AI. Let's take the next step together!**

aicopilot.com

ai_copilot

@aicopilot

ai_copilot

@aicopilot

# Workbook:
# Your AI Readiness Action Plan

This section is designed to help you reflect on your AI journey, assess your business's readiness, and create an actionable plan.

Use these prompts to document your thoughts, challenges, and next steps.

# Section 1:
## Understanding Your AI Goals

1. Define Your AI Vision

    - What are the top **three challenges** in your business that AI could help solve?

    **Challenge #1:** ______________________________

    ______________________________

    **Challenge #2:** ______________________________

    ______________________________

    **Challenge #3:** ______________________________

    ______________________________

    - What is your **primary goal** with AI implementation?

    ☐ **Automate repetitive tasks**

    ☐ **Improve customer experience**

    ☐ **Increase efficiency and reduce costs**

    ☐ **Gain a competitive edge**

    ☐ **Other:**______________________________

2. AI Knowledge Assessment

    - How would you rate your current understanding of AI?

    ☐ **Beginner (I know AI exists but haven't explored much)**

    ☐ **Intermediate (I've experimented with AI tools)**

    ☐ **Advanced (I actively use AI in my business)**

    - What is one AI tool or platform you are curious to explore further?

    ______________________________

    ______________________________

    ______________________________

    ______________________________

# Section 2:
## Evaluating Your Business Readiness

1. Your Data Strategy

- Do you currently collect and store business data in a structured way?

  ☐ **YES**

  ☐ **NO**

  ☐ **If no, what steps could you take to improve this?** ____________________

  ____________________

  ____________________

  ____________________

  ____________________

  ____________________

- What are the **top 3 data points** you believe could improve decision-making in your business?

  **1.** ____________________

  **2.** ____________________

  **3.** ____________________

2. AI and Team Collaboration

- Who in your business (or network) will be most involved in AI adoption?

  ____________________

  ____________________

  ____________________

  ____________________

- What is one skill gap in your team that may need to be addressed for AI adoption?

  ____________________

  ____________________

  ____________________

  ____________________

# Section 3:
## Identifying Your AI Champion

An **AI champion** is someone in your organization who:

- Embraces new technology and is open to learning.
- Can experiment with AI tools and help others adopt them.
- Understands the business's challenges and can align AI with company goals.
- Communicates effectively with both leadership and team members.
- Can advocate for AI-driven solutions while addressing team concerns.

Who in Your Business Could Be an **AI Champion?**

- Who in your business is naturally curious about technology?

___________________________________________

___________________________________________

___________________________________________

- Who has previously introduced a new tool, system, or workflow successfully?

___________________________________________

___________________________________________

___________________________________________

- Who regularly looks for ways to improve efficiency or reduce manual work?

___________________________________________

___________________________________________

___________________________________________

If you don't immediately have someone in mind, that's okay! You might be the best AI champion for now, or you can train and empower someone in your team.

What **Skills** Should an AI Champion Have?

- **Adaptability** – Open to change and learning.
- **Problem-Solving** – Can identify where AI might help.
- **Communication** – Can explain AI to team members.
- **Leadership** – Influences others to adopt AI.
- **Basic understanding of AI tools.**
- **Ability to research and test new technologies.**
- **Comfort with data analysis.**

Does your AI champion candidate have these skills?

- ☐ **Yes!** They already have most of these traits.
- ☐ **Somewhat.** They might need training in certain areas.
- ☐ **No.** I need to look for someone else or take on this role myself.

Who in Your Business Needs AI Training?

- **Customer Service Teams:** AI chatbots and automation can save time.
- **Marketing Teams:** AI can assist in content creation and ad optimization.
- **Operations & Admin Staff:** AI tools can automate reports, emails, and scheduling.
- **Sales Teams:** AI-powered insights can improve lead generation and follow-ups.
- **Leadership & Executives:** Understanding AI trends helps in strategic decision-making.

Who in your business would benefit from AI training?

________________________________________

________________________________________

________________________________________

________________________________________

________________________________________

# Section 4:
## Action Plan & Next Steps

An **AI champion** is someone in your organization who:

- Embraces new technology and is open to learning.
- Can experiment with AI tools and help others adopt them.
- Understands the business's challenges and can align AI with company goals.
- Communicates effectively with both leadership and team members.
- Can advocate for AI-driven solutions while addressing team concerns.

Who in Your Business Could Be an **AI Champion?**

- Who in your business is naturally curious about technology?

- Who has previously introduced a new tool, system, or workflow successfully?

- Who regularly looks for ways to improve efficiency or reduce manual work?

If you don't immediately have someone in mind, that's okay! You might be the best AI champion for now, or you can train and empower someone in your team.

# Final Thoughts

Take a moment to reflect on your responses. What **one insight** stood out to you the most from this workbook?

Now that you've mapped out your AI readiness, the next step is **taking action.** Start small, track your progress, and refine your approach as you go.

## Want help?

**Talk with our AI experts today**

**www.aicopilot.com**

Made in the USA
Monee, IL
05 November 2026

67985538R00045